Diana Lais Santos

Influence of Quality Raw Materials on the Brewing Process

AF301173

Diana Lais Santos

Influence of Quality Raw Materials on the Brewing Process

The importance of choosing ingredients

ScienciaScripts

Imprint

Any brand names and product names mentioned in this book are subject to trademark, brand or patent protection and are trademarks or registered trademarks of their respective holders. The use of brand names, product names, common names, trade names, product descriptions etc. even without a particular marking in this work is in no way to be construed to mean that such names may be regarded as unrestricted in respect of trademark and brand protection legislation and could thus be used by anyone.

Cover image: www.ingimage.com

This book is a translation from the original published under ISBN 978-613-9-66438-2.

Publisher:
Sciencia Scripts
is a trademark of
Dodo Books Indian Ocean Ltd. and OmniScriptum S.R.L publishing group

120 High Road, East Finchley, London, N2 9ED, United Kingdom
Str. Armeneasca 28/1, office 1, Chisinau MD-2012, Republic of Moldova, Europe
Printed at: see last page
ISBN: 978-620-8-03382-8

SUMMARY

INTRODUCTION

Beer is currently one of the most widely consumed alcoholic drinks in the world. According to the Ministry of Agriculture, Livestock and Supply - MAPA, Brazil is the third largest producer of beer in the world, producing an average of 14.1 billion litres/year, behind only China and the USA (CERVBRASIL, 2016, 2018).

Classified as a beverage produced by fermentation, with a low alcohol content (compared to other types of beverages) averaging between 3% and 8% alcohol by weight, beer is made from malted cereals, water and hops, fermented by yeasts. Beer has been shown to be an excellent source of soluble fibre and nutrients for a balanced diet (SILVA, 2005).

Because non-alcoholic beers are rich in vitamins and minerals and contain antioxidants, they are used as a source of rehydration during sports competitions (CERVBRASIL, 2018).

The diversity of this drink is not only restricted to taste and composition; the production process can also vary from one region to another. There is a variety of approximately 200 registered beers, considering only those with the best-known specifications (OETTERER, REGITANO- D'ARCE and SPOTO, 2006). The brewing process can be divided into 9 stages: mashing, filtration, boiling, clarification, cooling, fermentation, maturation and bottling (SILVA, 2008).

The daily consumption of beer, if done moderately, brings various health benefits. According to the Brazilian Beer Industry Association, men can consume up to 2 cans of beer a day and women up to 1 can (CERVBRASIL, 2017). Because it has anti-flammatory properties, beer helps to improve blood circulation in the brain, minimising the risk of suffering a stroke; it lowers LDL cholesterol (unhealthy cholesterol); it has a diuretic effect and can reduce the risk of developing kidney stones by 40%, due to the presence of phenolic compounds; Because it contains soluble fibre, it increases the amount of natural fibre in the body, regulating intestinal functions and reducing the chances of developing colon cancer and it is rich in B vitamins (B_2 , B_3 , B_5 , BI_2) (CERVBRASIL, 2017; SILVA, 2005).

In terms of per capita consumption, Brazil ranks 17th in the world. The country in first place is the Czech Republic, which consumes 143 litres per inhabitant per year, slightly more than a can a day (CERVBRASIL, 2018).

The brewing industry is the most important sector in the Brazilian economy, employing more than 2.7 million people and accounting for 1.6 per cent of GDP (CERVBRASIL, 2018).

Chemically, beer is a delicate and unstable beverage, undergoing various chemical and enzymatic reactions during its production and storage period, if carried out inappropriately.

With just four types of raw materials, it is possible to brew several varieties of beer. The time and manner in which each of these ingredients is added has a considerable influence on the type of drink produced. It is necessary to use quality raw materials and rigorous control in the brewing process so that there are no contaminating micro-organisms or alterations to the chemical and physical properties of the drink that negatively affect the final product.

CHAPTER 1

HISTORY: ORIGIN OF BEER

The first beers produced in history were made with barley, grapes, dates or honey. There are records of the use of beer by the peoples of Babylon, Sumeria and Egypt. There is no certainty about its real origin, but it is believed that the first brewing practices took place in Mesopotamia (now Iraq), where, as in Egypt, barley grows wild (SILVA, 2005).

There are indications that beer was discovered by chance. This occurred after a basket containing bread or a certain amount of grain was soaked by rainwater and then fermented by yeasts present in the environment, producing alcohol. And so the first beer in history was created (NIQUITO, 2016).

In Sumer around 6000 BC, it was women who were largely responsible for producing the drink, since the men were in charge of planting and harvesting, and it was they who produced food such as bread and flour, as well as beer. Malted barley grain was already used during this period (AMBEV, 2016; SILVA, 2005).

Around 1700 BC, brewing practices began to become so influential that in Babylon the drink was even used as a trading currency and used to define castes (relating people's daily consumption according to the social level they occupied). It is likely that the Babylonians first came into contact with beer through the Sumerians and then passed it on to the Egyptians. One of the first laws created about beer was the Code of Hammurabi, which defined criteria for brewing, consumption and commercialisation. One of the laws that made up the code determined the punishment for brewing poor quality beer: death by drowning in a barrel of one's own product (AMBEV, 2016).

Beer was the most widely consumed drink in Ancient Egypt. It was cultural for the population to consume beer during religious rites, and the drink was literally considered a divine creation. It has been reported that 2 million litres were distributed during the reign of Ramses III in 1225 BC. The Egyptians taught the Greeks the art of brewing, and the Greeks taught the Romans. Rudimentary beer was a darker, sweeter drink with less gas and a varied alcohol content (AMBEV, 2016) (AQUARONE; LIMA; ORZANI, 1993).

In the Middle Ages, the Catholic Church also began to get involved in beer production. The monks produced it from malted barley, boiled with hops and water and mixed with yeasts from the last fermentation. It was in the abbeys that beer began to be studied and perfected (CERVBRASIL, 2015) (MADRID; CENZANO; VICENTE, 1996).

In Germany, Duke William IV of Bavaria drew up the *Reinheitsgebot* Law, also known as the Beer Purity Law, which established criteria for the production of beer. By law, beer had to be produced using only barley malt, water and hops (yeast was not considered, as its existence was not yet known). What triggered the creation of this law was the rise in the price of wheat, because with the use of this cereal for beer production, the price of wheat began to rise, and the population was unable to buy it to eat (AMBEV, 2016).

In Brazil, the person responsible for creating the habit of drinking beer was Dom João VI, during the period when the Portuguese family was in the country, at the beginning of the 19th century, when the drink was imported from European countries. According to ROSA and AFONSO (2015) the Bohemia brand was the first beer to be produced in Brazil in the city of Petrópolis, RJ in 1853. In 1888 and 1891, Brahma and Antártica appeared, respectively. After more than 100 years of leading sales in Brazil, these last two brands merged in 2002 to form Ambev, the largest beer company in the country (BORZANI et al., 2008).

CHAPTER 2

DIFFERENT STYLES OF BEER

There are brewing schools, developed by great nations, in which it was possible through their technologies and production techniques to create various styles of beer that are still consumed today. Each of these schools produces its beers to specific standards, some of them more regulated and others more eccentric. There are four main schools: German, Belgian, British and American (NIQUITO, 2016);

• The German School created the beer purity law *(Reinheitsgebof)* in 1516 (SILVA, 2005). By law, beer must be produced using only barley malt, water and hops. The characteristics of their beers are more discreet, not very bitter and with a strong malt flavour. Examples include Weizen and Pilsen;

• The Belgian School gave rise to beers from the lambic family (spontaneous fermentation) and Trappist beers (produced by monks). Beers of varying styles were produced, but very aromatic, such as: acidic, spicy, fruity and even sweet, among them the Belgian Strong Golden, Dark Ale and Dubbel;

• The English School produced Ale-type beers and more bitter, malty and hoppy beers, the most common being the robust Porter, Stout and Indian Pale Ale;

• The American School is a mixture of the three schools, but in a more exaggerated and eccentric way, resulting in more complex beers such as Imperial IPA, American Pale Ale and Black.

Brazil does not have its own brewing school, but it is possible to find beers with characteristics from any of the four schools. Table I shows the types of beers mentioned in Brazilian legislation, including their country of origin, colour, alcohol content and type of fermentation. The beers considered to be of low fermentation are the Lager type and those of high fermentation are the Ale type, their differences being associated with the type of yeast chosen to make the drink.

TABLE I. Characteristics of the types of beer mentioned in Brazilian legislation.

Beer	Origin	Colouring	Alcohol content	Fermentation
Pilsen	Republic Czech Republic	Clara	4,1% a 6,0%	Low
Export	Germany	Clara	5,1% a 6,1%	Low
Lager	Germany	Clara	4,0% a 5,0%	Low
Dortmunder	Germany	Clara	5,1% a 6,1%	Low/High
Munchen	Germany	Dark	4,5% a 5,6%	Low
Bock	Germany	Dark	6,0% a 14,1%	Low
Malzbier	Germany	Dark	3,0% a 4,5%	Low
Stout	England	Dark	3,2% a 12%	Generally Low
Porter	England	Dark	3,8% a 6,0%	High or Low
Weissibier	Germany	Clara	4,3% a 5,6%	High
Ice	Canada	Clara	3,0% a 4,0%	-

(ADAPTED: SILVA, 2005)

The most widely consumed beer styles in Brazil, produced by the big breweries, are American Lager, which has characteristics that are best suited to the country's climate, being a light and refreshing drink, with an average alcohol content of 3.6% (in relation to its mass) and a neutral colour and bitterness. A very different drink compared to those produced by hand (OETTERER; REGITANO-D'ARCE; SPOTO, 2006).

There are Brazilian beers with unique flavours, using typically Brazilian ingredients such as rapadura, coffee, Brazil nuts, açaí and manioc (COLORADO, 2017).

CHAPTER 3

BEER COMPOSITION

Beer is a non-distilled alcoholic drink produced by the fermentation of beer wort by alcoholic yeasts. The wort consists of malted cereal (barley) boiled in good quality water, with the addition of hop flowers (Figure 1) (BRASIL, 2009).

Its composition includes esters, aldehydes, vicinal dicetones, organic acids, higher alcohols, phenols and iso-acids. The quality of the drink is related to the balance of volatile and non-volatile compounds (SIQUEIRA; BOLINI; MACEDO, 2008).

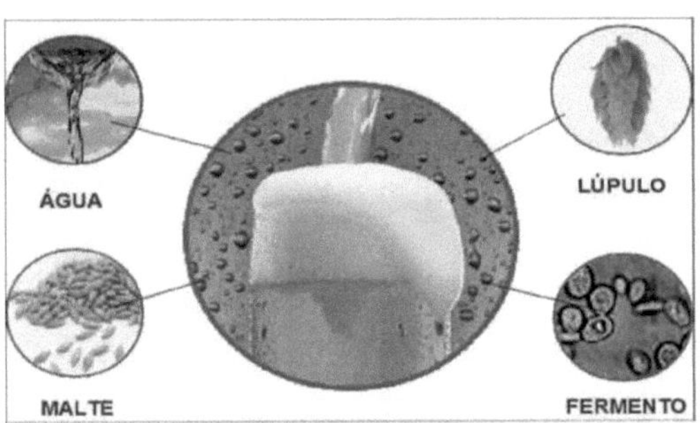

Figure 1 - Composition of beer

(SOURCE: ROSA and AFONSO, 2015)

Barley can be partially replaced by other cereals such as rice, wheat, rye, maize, oats and sorghum, whether malted or not (BORZANI et al., 2008).

Using just four ingredients, it is possible to produce countless styles of beer, ranging from flavours to textures and aromas. What makes this possible is the diversity found in these raw materials, the proportion used of each of them and the way in which the braising process is carried out.

To produce one of the best beers, the raw materials must be of the highest quality. The characteristics of each of the components will be shown below so that the process is efficient and the product is satisfactory.

3.1. Water

It is the main raw material for beer production, and approximately 95 per cent of its mass is made up of water. The volume of water used to produce beer is between 4 and 10 litres for every 1 litre of beer (SILVA, 2005).

Water found in nature, regardless of its source, has certain quantities of mineral salts and dissolved organic compounds. What defines quality water for brewing is the proportion of these salts and its pH. These compounds have a direct effect on the chemical and enzymatic processes involved in brewing and can generate unpleasant tastes and odours in the drink, directly affecting its quality (MADRID; CENZANO; VICENTE, 1996).

Nowadays, it is possible to buy equipment to treat the water, leaving it with the necessary characteristics for use in beer production, removing chlorine, adjusting mineral salts and removing undesirable solids. However, this equipment is expensive and is usually purchased by large breweries where it is used continuously (NIQUITO, 2016).

The high presence of chlorine in water can hinder fermentation and generate unpleasant odours in the drink due to the formation of chlorophenol. Yeasts naturally produce phenols and when they consume the chlorine in the water, they end up producing this compound (chlorophenol). This results in a beer with the *off-flavour* of medicine and antiseptics (NIQUITO, 2016).

3.1.1 Influence of water pH

A small proportion of water molecules (H_2O) have their bonds broken, generating hydroxyl OH' and hydrogen ions H^+. What defines pH is the concentration of these hydrogen ions in the solution. A neutral solution (pH = 7) has the same proportion of OH' and H^+, an acidic solution (pH < 7) has more H , while an alkaline solution (pH > 7) has a higher concentration of hydroxyls. For

In beer production, pH is defined by the hardness, salt concentrations and alkalinity of the water used (BJCP, 2012).

Water with an alkaline pH can dissolve undesirable compounds present in the malt husk. With a slightly acidic (6.5) to neutral (7.0) pH, amylolytic and proteolytic enzymatic reactions are favoured. The pH can be corrected using lactic acid or acidified malts. On the other hand, if the water is too acidic, the extraction of the compounds by the enzymes will be much lower, affecting the effectiveness of the process (AQUARONE; LIMA; BORZANI, 1993).

1.1.2 <u>Importance of salts and their concentrations</u>

For the production of certain beer styles, there will be water with specific characteristics, an important parameter being the degree of hardness of the water, defined by the amount of carbonated salts. Harder waters, with 200 to 300 mg/mL of calcium carbonates, are necessary for dark, sweet beers with malty flavours and a more defined body, such as Munich beers. And for lighter beers, waters containing calcium sulphate, 450 to 500 mg/mL (and a lower concentration of calcium carbonate), are better for lager-type beers such as Pilsen and lighter ales such as Burton (AQUARONE; LIMA; BORZANI, 1993) (MADRID; CENZANO; VICENTE 1996).

Each of the salts found in water has its importance in brewing. Calcium is the cation (positively charged ion) of greatest importance to the process, it reduces the pH of the wort to the acceptable range for production, preserves oxalate salts in solution (when precipitated they form turbidity) and helps in the coagulation of proteins. Magnesium serves as a nutrient for yeasts, and a concentration of 10-20 ppm is required (ppm = μm/mL or mg/L), as a quantity above this value causes an unpleasant mineral taste. Sodium, on the other hand, accentuates the sweetness of the drink in small quantities, but in higher concentrations gives a salty flavour (BJCP, 2012).

Bicarbonate (HCO_3) is an anion (negatively charged ion) that determines the basicity of water. It neutralises the acidity caused by roasted malts (as in Porters, Stouts and Dunkels), reduces the hardness of the water by reacting with calcium and causes the extraction of compounds such as tannin, responsible for giving the beer its mouthfeel. The sulphates SO_4^{-2} accentuate the bitterness of the hops and the chlorides Cl^- in low quantities, enhance the sugar, but make it difficult for the yeasts to flocculate if they are in high quantities (BJCP, 2012).

1.1.3 <u>Basic production requirements</u>

Other basic water requirements to obtain an optimum quality end product should be: pH between 6.5 and 7.0 to favour enzymatic action, in which starch is reduced to fermentable sugar. Magnesium values of less than 100 mg/litre, traces of magnesium sulphate, 200 to 300 mg/litre in sodium chloride, and iron concentration of less than 1 mg/litre (AQUARONE; LIMA; BORZANI, 1993).

1.2. Malt

Malt is cereal grain that has been germinated, dried and roasted in order to increase the percentage of sugar in the grain and to form the protease and amylase enzymes needed for the fermentation stage. The malt used in breweries comes from barley, but corn, wheat, oats, rice, rye and various other cereals can be malted. However, barley is the one that presents the fewest technical difficulties during malting. For example, corn has problems with rancidity in the grain (decomposition caused by contact with the air, resulting in an unpleasant odour and sour taste) and wheat suffers from the appearance of microorganisms on its surface (BORZANI et al., 2008). Figure 2 shows the types of barley malt that exist.

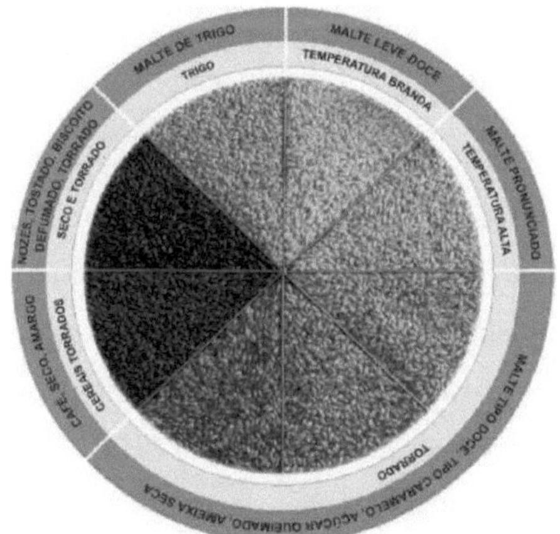

Figura 2 - Types of malt.

(SOURCE: CERVBRASIL, 2017)

Malt can be found in three forms: "Green" Malt, which is malted cereal grains;

Malt Extract, prepared by immersing ground malt in hot water; and finally Malt Syrup, the concentration of the extract so that the enzymes are not lost (AQUARONE; LIMA; BORZANI, 1993).

The enzymes α-amylase, β-amylase, maltase and protease are activated in the malt during the malting of the grain and act in the wort, transforming the starches into fermentable sugars (BOBBIO; BOBBIO, 2003).

The starch in malt grain has shorter chains than that found in barley and is more soluble and softer. The amount of sucrose, if compared, is 0.5 to 1.0% in barley to 8.0 to 10.0% in malt and its diastatic power (high enzymatic activity) is around 50% in barley and 250% in malt (SILVA, 2005) (OETTERER; REGITANO-D'ARCE: SPOTO, 2006).The table below shows the average composition of malt compared to barley grain:

Table II - Composition of barley grain and malt.

Features	Barley	Malt
Grain mass (mg)	32 a 36	29 a 33
Humidity (%)	10 a 14	4 a 6
Starch (%)	55 a 60	50 a 55
Sugars (%)	0,5 a 1,0	8 a 10
Total nitrogen (%)	1,8 a 2,3	1,8 a 2,3
Soluble nitrogen (% of total N)	10 a 12	35 a 50
Diasporic power, Lintner	50 a 60	100 a 250
a-amylase, dextrin units	Traits	30 a 60
Proteolytic Activity	Traits	15 a 30

(SOURCE: SILVA, 2005)

3.2.1 Barley

Barley is a grass of the genus *Hordeum vulgare L.,* its grains are in the form of ears, in two or six rows. Barley grains can be distinguished by the number of fertile flowers; two-row barley has two of the six rows of fertile flowers that can produce grains, while six-row barley has all of its rows with these qualities (BJCP, 2012).

If you compare them, six-row barley has less starch, higher enzymatic activity (protein richness, a great option when using large quantities of adjuncts in beer formulation) and more hulls, so it performs better in filtration, while two-row barley has a higher mashing yield (the size of its grains is larger) and is less difficult to grind (SILVA, 2005) (BORZANI, 2008).

Barley grain is made up of a husk on the outside and starchy endosperm and the embryo, also known as the germ, on the inside (BORZANI, 2008).

The husk is made up of a cellulosic material and contains proteins, resins and tannins (in smaller quantities). It is used in the filtration stage, where it is very important as a filtering element for the must. The endosperm is the inner part of the grain, which accumulates starch that will be converted into fermentable sugar by the action of enzymes (it feeds the embryo during development).
germination) and will serve as food for the yeasts during the fermentation stage and the germ is the region that will start the germination process, this occurs in the presence of air and water (BORZANI, 2008).

3.2. 2Malting

The purpose of malting the grain is to convert the huge starch chains into soluble starches and to activate the enzymes that will reduce the proteins and sugars into suitable components at the mashing stage. These are: amylase enzymes (α-amylase, β-amylase), glucanases, proteases and debranching enzymes (BJCP, 2012).

The malting process consists of three stages: maceration, germination and drying. The first stage consists of placing the grains in favourable conditions to induce germination. During this period, temperature, humidity and aeration controls must be carried out (SILVA, 2005). Firstly, the grains are cleaned and placed in cylindrical tanks to be macerated, the water is changed every 6-8 hours and the temperature is maintained at between 5 and 18°C. After two days, this stage is finished. During this period the barley has reached 42 to 48 per cent humidity and the radicle appears (BORZANI, 2008).

They are then left to germinate at a temperature of 15 to 21°C with a constant injection of oxygen for a period of 3 to 4 days. At this stage, the development of the embryo is observed and the change in the texture of the grain and the odour released are monitored. When the structure of the embryo reaches two thirds of the size of the grain, the stage is finished (BORZANI, 2008).

The final stage is the drying of the grains, using hot, dry air to stop the germination process and provide the malt with specific colour and aromas. If drying is carried out over a long period of time and at low temperatures, a clear malt with high enzymatic power will be obtained, but if it is done quickly and hotly, dark malt will be produced with poor enzymatic content (BORZANI, 2008).

The malt should be kept to rest for approximately one month before being used for the first time (BJCP, 2012).

3.3 Hops

Humulus lupulus, known as hops, is a hermaphrodite plant (it has female and male flowers) belonging to the *Cannabinaceae* family, but it doesn't have the hallucinogenic action of *Cannabis.* The unfertilised female flowers are used because their aromatic power is higher (BORZANI et al., 2005) (OETTERER; REGITANO-D'ARCE; SPOTO, 2006).

Hops bring a number of benefits to beer: they act as a "spice", providing bitter flavour and aromas to the beer. It is also important in the generation of foam and helps maintain it; it has antiseptic properties, acting as a natural preservative in the drink (MADRID; CENZANO; VICENTE, 1996).

3.3.1 Hop composition

The female flowers have glands that contain lupulins, bitter resins and essential oils. The resins are mainly composed of alpha- and beta-acids, the main source of bitterness. Essential oils are composed of hydrocarbons from the terpene family, esters, aldehydes, ketones, acids and alcohols, influencing beer styles and generating characteristic flavours and aromas (BORZANI et al, 2005).

The compounds in hops that provide the specific flavour in the drink vary according to the species of each plant. The proportion of each compound is approximately 0.2 to 3% essential oils, 1.5 to 9.5% β-acids (known as lupulones) and 2.0 to 16% α-acids (called humulones) (SILVA, 2005).

Essential oils provide the aromatic character of hops in beer, and their composition is very complex, with more than 200 compounds. After undergoing isomerisation, a-acids are responsible for the main bitterness in beer, providing around 70% of its bitter taste, which is more intense than the original, non-isomerised acids (isomers: substances that have the same molecular formula, but differ in structural formula). On the other hand, β-acids don't have much bitterness, but they do have great bactericidal power over Gram-positive bacteria, acting on the cell membrane with the transport of metabolites and altering the pH inside the cell, inhibiting the growth of bacteria (SILVA; FARIA, 2008).

Table III - Chemical composition of flowering hops.

FEATURES	PERCENTAGE (%)
Total bitter resins	12-22
Proteins	13-18
Cellulose	10-17
Polyphenols	4-14
Humidity	10-12
Mineral salts	7-10
Sugars	2-4
Lipids	2,5 - 3,0
Essential oils	0,5-2,0
Amino acids	0,1-0,2

(SOURCE: SILVA, 2005)

3.3.2 Cultivation   and marketing

Hops are grown in regions where the summer light periods are longer, approximately 15 to 18 hours a day, such as regions between latitudes 35° and 55°, and the harvesting period usually takes place between mid-August and mid-September (ABOUMRAD; BARCELLOS, 2015).

It is commercialised in various forms, such as pellets, extract, powder and dried flowers. Some of the best hops are imported from West Germany, in the region of Bavaria (AQUARONE; LIMA; BORZANI, 1993). New products are being created thanks to the advent of technology, and can already be found in the shops, such as isomerised extracts, which can be used after the end of fermentation in order to adjust the bitterness. Another type of hop extract is isomerised and reduced, which makes it possible to increase foam retention in the drink and helps protect against light. The use of these extracts together and/or individually will depend on the type of beer to be brewed and the needs of each situation (SILVA, 2005).

Figure 3 - Hops.

(SOURCE: AMBEV, 2016)

3.4. Adjuncts

These are products containing fermentable carbohydrates (sugar that will be metabolised by the yeast) used to replace malt, with the intention of reducing costs, which are added during the preparation of the wort, for example, cereals such as corn, wheat and rice, which can be malted or not (AQUARONE;LIMA; BORZANI, 1993).

As mentioned above, barley malt has a high diastatic power, which is of great importance when using other non-malted starch ingredients. The enzymes in the malt are responsible for breaking down the starch present in the adjuncts. However, adjuncts in the form of syrup and sugars do not need to go through the saccharification process (D'AVILLA et al., 2012).

Syrups derived from barley, corn and wheat can be used. And sugars such as sucrose (derived from sugar cane) or invert sugar, hydrolysed sucrose (BORZANI et al., 2008).

It is possible to brew beer with unmalted barley. You should have a maximum proportion of 20% of the total grain used, and unmalted cereals have lower nitrogen contents, leading to a loss of foam stability. However, greater foam stability was found in beers produced with malt and barley than those containing only malt, probably due to the lower level of protein degradation by enzymes. A negative aspect of beers containing both ingredients (malt and barley) is a harsh flavour in the final product, interfering with the sensory quality of the product. As such, their use is not highly recommended (D'AVILLA et al., 2012).

Another type of additive widely used by large breweries are products whose purpose is not to change the composition of the beer or its flavour, but only to contribute to its better appearance and stability. These are: flocculators (prevent cloudiness due to impurities), colourants, antioxidants (remove some proteins from the wort that could cause cloudiness), foam stabilisers (help maintain the foam, which is of great importance for maintaining the flavour of the beer, preventing carbon dioxide from escaping and oxidising the drink during consumption) (ABOUMRAD; BARCELLOS, 2015).

3.5. Yeast

Yeast is defined as a unicellular fungus, an extremely important microorganism used in the food industry. The species used for beer production is from the *Saccharomyces* family, *which* is a mixture of different strains and has intense fermentative activity. It is responsible for metabolising the sugar present in the wort and producing alcohol, carbon dioxide and aromatic compounds (FRANCO; LANDGRAF, 2005). During fermentation, between 600 and 800 flavour components are produced, causing major changes in the taste and aroma of the drink (ABOUMRAD; BARCELLOS, 2015).

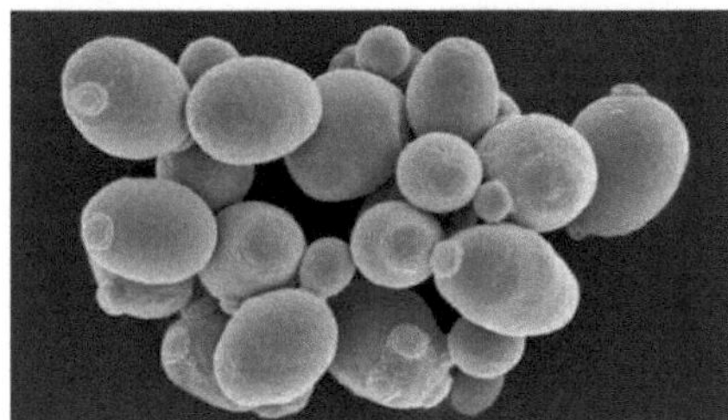

Figura 4   - *S. cerevisiae* yeast

(SOURCE: FRANCO and LANDGRAF, 2005.)

Pure yeast strains are used at the start of fermentation. They are selected according to the type of beer to be produced. *S. cerevisiae* is used for high-fermentation ale-type beers (Figure 3) and *S. carlsbergensis* or *uvarum* is used for low-fermentation lager-type beers. Normally the strains are selected after genetic analysis, separating out those that best suit the manufacturing process, those that ferment more efficiently and those that are best adapted to the raw materials and processes and the consumer's taste preferences for the style of the drink (AQUARONE; LIMA; BORZANI, 1993).

TABLE IV - Change in nomenclature for *S. cerevisiae*

1952	1970	1984
S. cerevisiae	*S. cerevisiae*	
S. willianus		
S. coreanus	*S. coreanus*	
S. carlsbergensis		
S. uvarium	*S. uvarium*	
S. logos		
S. bayanus	*S. bayanus*	
S. pastorianus		
S. oviformis		
S. beticus		
S. heterogenicus	*S. heterogenicus*	
S. chevalieri	*S. chevalieri*	*S. cerevisiae*
S. fructuum		
S. italicus	*S. italicus*	
S. steineri		
S. globosus	*S. globosus*	
	S. aceti	
	S. diastaticus	
	S. oleaginosus	
	S. prostoserdovil	
	S. capensis	
	S. inusitatus	
	S. hispalensis	
	S. cerevisiae	

SOURCE: SILVA, 2005

3.5.1 Reserve carbohydrates   in yeast

At the end of fermentation, the carbohydrates glycogen and trehalose are accumulated by the yeast cells; these two compounds are produced by the microorganism itself. Glycogen is made up of several D-glucose chains with α-(l,4) bonds. These chains have around 12 D-glucose residues and their branches are formed by α-(l,6) bonds. The disaccharide trehalose is a carbohydrate made up of two D-glucose residues, making bonds through their reducing carbons (SILVA, 2005).

Glycogen provides energy for cells during the period when lipids are being synthesised, during the aerobic fermentative stage (presence of oxygen) and serves as a biochemical reserve source of energy and carbon for cells during periods of starvation. Trehalose, on the other hand, protects the plasma membrane from

yeast against their spontaneous self-destruction (autolysis), guaranteeing their viability during germination, starvation and dehydration (SILVA, 2005).

3.5.   2Contaminating agents

Other types of yeast can develop during the fermentation process and are considered a dangerous infection because they directly affect the quality of the beer. These are called wild yeasts (yeasts other than those cultivated, which have not been added to the process), such as other species of *Saccharomyces* and yeasts of the genera *Schizosaccacharomyces, Hansenula, Candida, Pichia, Brettanomyces, Kloeckera* and *Torulopsis*. They are involved in spoiling the drink, producing turbidity, weakening fermentation and can generate abnormal flavours and aromas. To prevent these contaminating yeasts and to detect bacteria, routine microbial analyses should be carried out, thus guaranteeing the quality of the yeast (AQUARONE; LIMA; BORZANI, 1993) (BORZANI, 2008).

Like wild yeasts, contaminating bacteria affect both the wort and the final product of the process, the beer. They cause turbidity, unpleasant flavours and odours. One way of controlling bacteria would be to wash the yeast with mineral acids, with a pH close to 2.5, for approximately two hours. Or discard the yeast strains after five to ten batches of brewing (BORZANI, 2008).

Contamination of the must can occur due to a lack of asepsis in the equipment and hygiene on the part of the operator, or by keeping the pH low indefinitely, creating favourable environments for its development. Cleaning can be done by applying jets of water (under pressure) to remove the thickest dirt on the walls of the equipment; treatment with a heated solution of caustic soda with sodium hypochlorite, as the soda acts as a disinfectant (germicide) and detergent (dissolves proteins) and the chlorine in the hypochlorite acts as a bactericide (BORZANI, 2008).

CHAPTER 4

PRODUCTION PROCESS

The beer production process is quite simple, but strict control must be maintained so that problems don't occur during the manufacturing stages, generating chemical and physical alterations in the drink and impacting on its final quality (AQUARONE, LIMA and BORZANI, 1993).

Processes can be broken down into several stages. Figure 4 shows a flowchart of the production process, according to Silva, 2008:

Malt milling→ Roasting or mashing→ Filtration→ Boiling→ Cooling→ Fermentation→ Maturation→ Carbonation→ Bottling→ Pasteurisation.

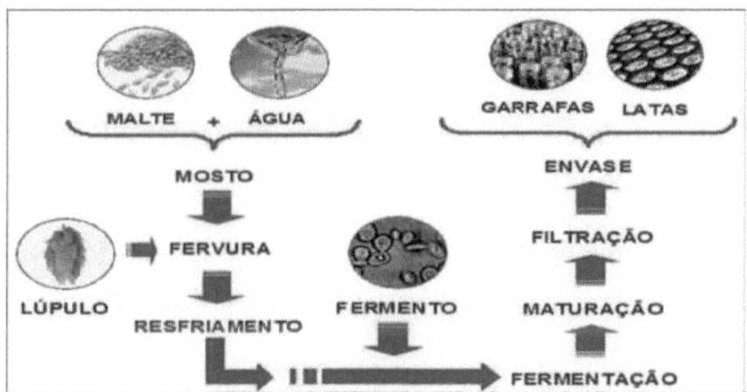

Figura 5 - Summary of the Brewing Process

(SOURCE: ROSA; AFONSO, 2015)

The malts are milled so that the husks are opened from one end to the other and the inside of the grain is fragmented (but not to the point of becoming powder). The production process begins with mashing (also known as mashing). The sugar is extracted from the malt by breaking down the starch through the action of its natural enzymes. The wort is then filtered and boiled (hops are added at this stage) (SILVA, 2005).

By centrifuging, natural decantation is carried out, removing the suspended particles that give the liquid a cloudy appearance, resulting in crystal-clear beers. The wort is quickly cooled to the right fermentation temperature. The yeasts are added and fermentation begins, lasting approximately seven days. The yeast transforms the sugar into alcohol and carbon dioxide gas (SILVA, 2005).

During maturation, the beer "rests" at low temperatures, improving its aroma and flavour, which can take up to 21 days. Secondary fermentation takes place and the drink is lightly carbonated due to the action of the yeasts. The beer is bottled and pasteurised (in the case of beers, Chopp doesn't go through this stage). It is then ready for sale (ROSA; AFONSO, 2015).

All the manufacturing processes have a direct impact on the quality of the end product. The importance of each of the stages will be discussed, describing the reactions involved in each one:

4.1 Mixing or mashing

At this stage the brewing wort is formed, consisting of crushed malt and water. If necessary, some adjunct is added, such as syrup or malt extract. The purpose of mashing is to extract the total solids from the malt and then hydrolyse them. Hydrolysis consists of breaking down the starch and amylose molecules into simpler carbohydrates (AQUARONE; LIMA; BORZANI, 1993).These simpler carbohydrates are, for example, the monosaccharide glucose ($C\,H\,O_{6126}$) and the disaccharide maltose ($C\,H\,O_{122211}$), (made up of 2 glucose molecules), both shown in figure 6, and the third carbohydrate is the trisaccharide maltotriose $C\,H\,O_{183216}$, made up of 3 glucose molecules. The efficiency of this stage is the result of the right temperature, pH, mash concentration, stirring and mashing time (AQUARONE; LIMA; BORZANI, 1993).

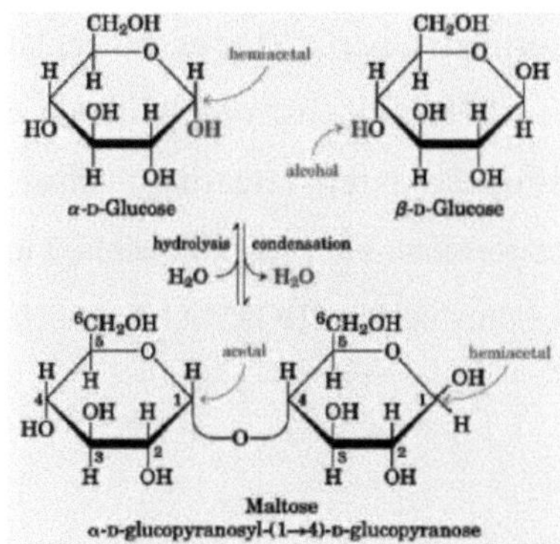

Figura 6 - Representation of the carbohydrates Glucose and Maltose

(SOURCE: GUEIROS, 2015)

Starch (C H O_{6105n})(made up of glucose molecules) is made up of two polysaccharides, 24% amylose (a linear polymer made up of glucose units with α-1.4 bonds) and 76% amylopectin, shown in figure 7, (a branched polymer made up of glucose, also with α-1.4 bonds, but with α-1.6 bonds at the branching points). Amylose is the soluble part of starch, also known as the saccharifier, and because it is easily fermentable it is transformed into fermentable sugar. Through genetic improvement of the cereal grain, levels of up to 47 per cent amylose can be achieved (OETTERER; REGITANO- D'ARCE; SPOTO, 2006).

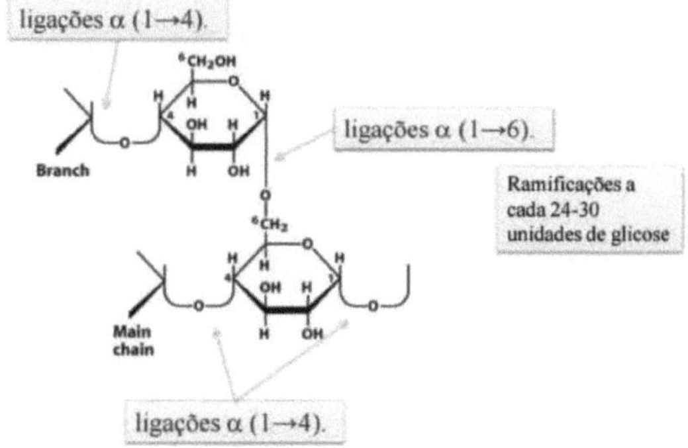

Figura 7 - Representation of an Amylopectin molecule

(SOURCE: MIYAMOTO, 2011)

The a-amylase enzymes act on amylopectin, forming dextrins (lower molecular weight polysaccharides), such as zso-maltose, made up of two glucose units with α-1,6 bonds. The β-amylases act on the non-reducing part of the starch, releasing maltose, glucose and maltotriose (Figure 7 and Figure 8). The fermentable sugars generated in this process will be converted into alcohol and carbon dioxide (BORZANI, 2008). (BOBBIO; BOBBIO, 2003).

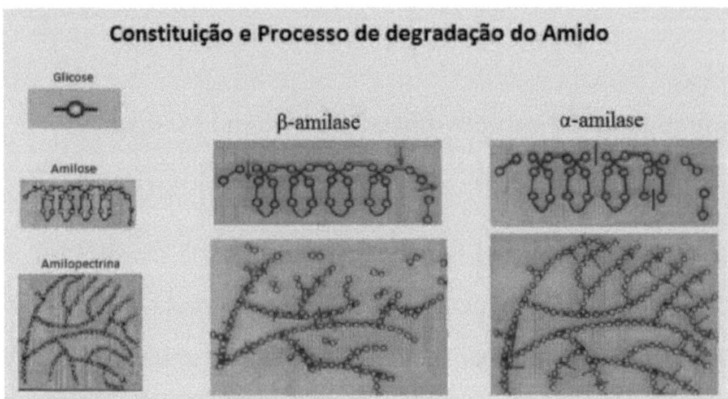

Figura 8 - Starch degradation process
(SOURCE: NIQUITO, 2016)

The speed at which the molecule undergoes hydrolysis depends on the geometry of the sugar chain, which is influenced by the presence of branches. The bonds in linear chains are hydrolysed more easily than those that give rise to branches, however, the hydrolysis of branched oligosaccharides takes place more quickly than that of maltose and maltotriose (FILHO VENTURINI, 2005).

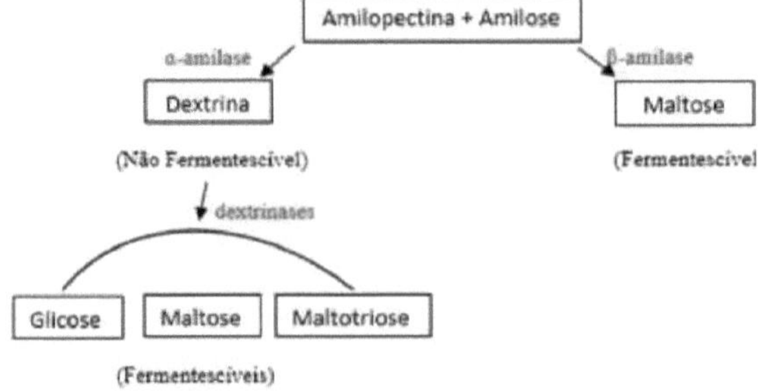

Figura 9 Starch Degradation Process

Each enzyme acts at a different temperature, also depending on the type of beer to be brewed (BORZANI, 2008).

The process in which starch and amylose are transformed into smaller sugars is called saccharification. The result of this breakdown is directly linked to fermentation, carried out by yeasts, which will define the beer's body and alcohol content. In an efficient fermentation, the end product is a weaker beer, with less body and more alcohol. On the other hand, if it contains more non-fermentable sugars (longer chain sugars that the yeast can't metabolise), you get a sweeter, fuller-bodied drink with a lower alcohol content (BORZANI, 2008).

When using protein-rich malts such as wheat and oats, the initial mash temperature should be between 50°C and 55°C, so that the activity of proteolytic enzymes, proteases and peptidases, is favoured, generating the breakdown of protein peptide bonds, which are long chains of amino acids and forming chains.

such as amino acids and peptides (2 or more amino acids) (BORZANI, 2008).

The function of proteins is to catalyse chemical reactions involving enzymes (BOBBIO; BOBBIO, 2003).

The enzymes α-amylase, β-amylase, maltase and protease are activated in malt during the malting of the grain and act on the wort, which is made up of 85% to 90% fermentable carbohydrates. Each enzyme acts at a different temperature, also depending on the type of beer to be brewed (BORZANI, 2008). The table below shows the temperature and pH ranges of the enzymes present in malt and their function in the brewing process:

Table V - Temperature and pH at which the enzymes work

Enzyme	Optimum temperature range	Range pH work	Function
Phytase	30 - 52,2°C	5,0-5,5	Reduces the pH of the must. No

			longer used
Shredder	35 -45°C	5,0-5,8	Solubilisation of starches
β-Glucanase	35 -45°C	4,5 - 5,5	Improves gum breakage
Peptidase	45 - 55 °C	4,6-5,3	Produces Arnino Free Nitrogen
Protease	45 - 55°C	4,6 - 5,3	Breaks down large proteins that cause turbidity
β-Amylase	55-66,1°C	5,0-5,5	Produces maltose
α-Amylase	67,8- 72,2°C	5,3 - 5,7	Produces a variety of sugars, including maltose

(SOURCE: BORZANI, 2008).

This *is* followed by an iodine solution test, which confirms the complete hydrolysis of the starch. This last process is followed by a final increase in temperature, to 76°C to 80°C, in order to inactivate the enzymes in the must. E

It is important not to exceed 80°C so that phenolic compounds, which promote an astringent taste, are not formed (AQUARONE; LIMA; BORZANI, 1993).

4.2 Filtration

At the end of the mashing process, the filtration process begins, which consists of separating the malt extract dissolved in water (wort) from the insoluble residue of the malted cereal (husks or bagasse), which settles at the bottom of the filtration trough (ROSA; AFONSO, 2015).

The malt husks serve as a natural filter for the wort. By gravity, the wort passes through the bagasse, initiating the first stage of filtration. The second part consists of washing this layer formed by the husks with water at 75°C, bleaching out the remaining sugar extraction, resulting in minimal waste of raw materials (SILVA, 2005).

Water at 75°C is used because the temperature influences the viscosity of the wort, favouring the complete separation of the two phases that make up the mixture, obtaining a wort with minimal turbidity (a very clear liquid free of solid particles), indicating successful filtration, and there is no possibility of extracting insoluble substances, such as tannins, from the malt husk (BORZANI, 2008).

In Brazil, one piece of equipment that is widely used for filtration is the filtration vat: it is made of stainless steel and has a false bottom like a sieve, but you can use a combined mash-filtration vat or a filter press (PICCINI; MORESCO; MUNHOS, 2002) (BORZANI, 2008).

When choosing the best filtration method, aspects such as ease of cleaning, space occupied by the equipment, cost-effectiveness and speed of the process must be taken into account (PICCINI; MORESCO; MUNHOS, 2002).

4.3 Boiling

At this stage the wort is boiled with the intention of inactivating the enzymes that cause the proteins to coagulate and guaranteeing their biological and biochemical stability. During this stage, hops are added to the wort, which are responsible for giving the drink its characteristic aroma and flavour (SILVA, 2008).

During boiling, several important reactions take place, such as the total inactivation of the alpha-amylase enzyme, the formation of "trub", a gelatinous sludge-like residue made up of coagulated proteins and tannins that will later be removed by a second filtration, the microbial flora that resisted the previous processes is destroyed and some sugars are caramelised (BORZANI, 2008).

The boiling process can last from 60 to 90 minutes, with a loss of up to 10 per cent of the initial volume due to evaporation. Hops can be added at the beginning or end of the stage, in concentrations of 0.4 to 1.4 g/L in relation to the initial volume, and the amount added to the wort will depend on the style of beer being produced (BORZANI, 2008) (PICCINI; MORESCO; MUNHOS, 2002).

9.4 Fermentation

One of the most important stages in beer production is fermentation. In this process, the starch present in cereals is converted into alcohol by yeasts. Such is the importance of this stage that one of the ways of characterising types of beer is by the type of fermentation. There are three types:

- Lager beer: beers produced by low fermentation, where the yeast is deposited at the bottom of the fermenter. These are beers with a lower alcohol content and greater lightness. For the process, the temperature needs to be between 12°C and

14°C, taking twelve days to complete fermentation. Examples: Pilsen, Bock, Malzbier (NIQUITO, 2015). Yeasts continue to work even at temperatures close to 0°C, affecting the reduction of other fermentation products, resulting in a final product with a softer, purer flavour. This variety of beer, lagers, can be considered a landmark of the contemporary age and began to be produced for commercial purposes after the arrival of refrigeration equipment in the 19th century (BJCP, 2012).

- Ale beers are brewed by top fermentation. The yeasts remain on the surface of the wort and the process takes seven days to complete at temperatures of 19°C to 21°C. Ale beers generally have a higher alcohol content, are fuller-bodied and have a strong flavour. Examples: Stout, India Pale Ale (IPA), Weizenbier (Wheat Beer) (NIQUITO, 2015). Their yeasts are more sensitive than those of the lager type; when left to mature for a long time at a very low temperature, the cells flocculate and become inactive (BJCP, 2012).

- Lambic beer: a rustic way of brewing beer, fermentation takes place spontaneously by wild yeasts (yeast that is not selected) present in the environment. This type of beer is produced in Europe, around Brussels, where the region has an ideal microflora for brewing this beer. Examples: Trappists.

Secondary products are esters (ethyl acetate (C H O_{482}), isoamyl acetate (C H O_{7142}), n-propyl acetate (C H O_{5102})), acetic acid (CH_3 COOH), propionic acid (C H O_{362}) and higher alcohols (1-propanol, 2-methyl-1-propanol, 2-methyl-1-butanol and 3-methyl-1-butanol). The secondary products are responsible for giving beer its physical and sensory characteristics (ROSA; AFONSO, 2015).

At the start of the process, the must with yeast is left in the presence of oxygen. The purpose is to generate the reproduction and invigoration of the yeasts (generating more yeasts), and for the production of energy, represented by the following equation (BORZANI, 2008):

$$C_6H_{12}O_6 + 6\ O_2 \rightarrow 6\ CO_2 + 6\ H_2O + 38\ ATP + calor$$

To carry out alcoholic fermentation, more than a dozen enzymes are involved. According to Borzani (2008), this oxidation-reduction process is represented by the following equation:

$$C_6H_{12}O_6 \rightarrow 2\ C_2H_5OH + 2\ CO_2 + 2\ ATP + calor$$

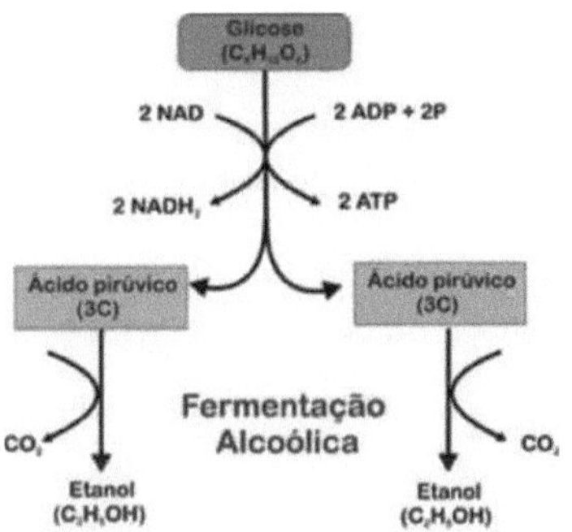

Figura 10   Alcoholic Fermentation Equation
(SOURCE: ABOUMRAD; BARCELLOS, 2015)

During the fermentation process, yeasts metabolise glucose to obtain energy (ATP). The unit of glucose broken down is converted into 2 atoms of pyruvic acid, or pyruvate, and energy, by the process of glycolysis. Pyruvate is then transformed into ethanol (C H_{25} OH) and carbon dioxide (CO_2), as shown in figure 10 (ABOUMRAD; BARCELLOS, 2015).

As the fermentation process releases heat, it is necessary for fermentation domes to contain cooling equipment so that the temperature remains constant (SANTOS, 2003).

Depending on the genetics of each type of yeast, whether the flocculation process takes place late or early will depend on the composition of the wort. Lager yeasts will flocculate and form clusters that are deposited at the bottom of the dome at the end of primary (or main) fermentation, which can take seven to ten days. High fermentation yeasts, on the other hand, will have their cells adsorbed in the *CO* gases$_2$ and will be carried to the surface of the mixture, where they will be collected at the end of the process, indicating the production of ale-type beer (AQUARONE; LIMA; BORZANI, 1993) (SILVA, 2005).

4.5 Ripening

The method of maturation will depend on the type of yeast used, normally a temperature of 0°C is used for a period of one to four weeks. The beer needs to have a fermentable extract concentration of 0.5-1.5% m/m (mass of solute by mass of solution) and approximately 2.0×10^6 to 5.0×10^6 cells /mL of viable yeast (BORZANI, 2008).

At this stage, small and subtle transformations take place, one of which is secondary fermentation, where the residual carbohydrates continue to be slowly fermented by the remaining yeasts, and the beer is saturated with CO_2 . These same yeasts are responsible for the chemical changes that help improve the flavour of the drink, transforming: acetaldehyde ($C_2 H4O$) into acetic acid (CH3COOH), vicinal dicetones such as 2,3-pentanedione ($CH_3 COCOCH_2 CH_3$) into 2,3-butanediol ((CH $)_{32}$ $(CHOH)_2$), and sulphur compounds such as diethyl sulphide, ($C H)_{252} S$, in ethanol ($C H_{25} OH$) and inorganic sulphates (ROSA; AFONSO, 2015).

The production of carbon dioxide during the fermentation of the remaining extract causes the beverage to carbonate to an ideal concentration for beer; if this level is not reached, it can be corrected after the filtration phase (SILVA, 2005).

The low temperature allows the remaining yeasts, unstable proteins and resins to precipitate, thus clarifying the drink (AQUARONE; LIMA; BORZANI, 1993).

The flavour and aroma of mature beer are generated by the formation of esters, predominantly ethyl acetate with a concentration of 21.4 mg/L and amyl acetate 2.6 mg/L (AQUARONE; LIMA; BORZANI, 1993).

In the maturation stage, additives can be used to improve foam, adjust colour, odour and flavour, guarantee the beer's stability against cloudiness and prevent the development of infections (BORZANI, 2008). At the end of this phase, the beer is practically complete with a defined final aroma and flavour.

4.6 Clarification

At the end of maturation, it is possible to detect the occurrence of turbidity in the beverage, which makes it necessary to clarify it through filtration. This doesn't change the flavour or composition of the beer, but it is very important to give it a crystal-clear appearance. Filtration can be done using a diatomaceous earth filter and then passed through a plate filter or cartridge filter. It is now possible to find membrane filters available on the market which, as well as filtering, sterilise the product, so that the pasteurisation stage can be dispensed with and the beer can then be bottled (SILVA, 2005).

Regardless of the type of filter used, great care must be taken when filtering beer to avoid the loss of CO_2, the entry of oxygen and the contamination of the product with microbiological organisms (BJCP, 2012).

4.7 Pasteurisation

Pasteurised beer is a drink that has undergone a thermal process in which it has been subjected to a high temperature and then cooled. Pasteurisation aims to confer biological stability by destroying the micro-organisms that spoil beer (SILVA, 2005).

There are two ways of carrying out pasteurisation: the first consists of using modified plate heat exchangers, and the second option would be a pasteurisation tunnel. In the case of using heat exchangers, the process must be carried out before the beverage is filled into bottles, cans or kegs. The practice is based on keeping the beer for a few seconds at 75°C with a pressure of 7.5-10 bar at the inlet and 1-5 bar at the outlet of the exchanger, after which it is cooled. The second option would be to pass the canned or bottled beverage through a pasteuriser tunnel, spraying hot water through all sections of the equipment, inside the bottles temperatures reach around 60-65°C, cooling consists of spraying with cold water (BORZANI, 2008).

CHAPTER 5

SENSORY ANALYSIS

The organoleptic characteristics of the great beers from every region of the planet were produced from a simple blend of tradition, invention and popular demand. It was after 1840, with the arrival of the industrial and scientific revolutions, that the interpretation of the form of production slowly changed. Before that, beers were produced without industrial procedures and with no ambition to transport them long distances (SILVA, 2005).

In the last half century, there has been a significant increase in research into the sensory characteristics of beer, such as aroma and flavour. These two characteristics are the basis of the drink's success with consumers, so it is necessary to have greater knowledge on the subject, knowing and controlling them.

According to Borzani (2008), the main defects associated with taste and/or odour that can be found in the drink are presented below

5.1 Turbidity

There are two possible causes of turbidity in beer. The first would be the growth of yeasts or bacteria in the unpasteurised beverage or improper storage (refrigeration).

The second factor would be the coagulation of colloids, for example when the concentration of calcium oxalate is higher than 20 ppm (20 mg/L) or due to the presence of polysaccharides (carbohydrates), occurring when low quality raw materials are used and when there is alternating temperature (cooling/heating).

One way to reduce the possibility of turbidity would be to: increase the amount of adjunct in the mash; use malts with high diastatic power; boil the wort vigorously and allow aeration; use protein rest in the mashing stage; store the bottled drink at the lowest possible temperature (without freezing) and avoid too much agitation, temperature variation and the incidence of light.

5.2 Insipidity

It consists of a lack of carbonation and foam. The lack of carbonation can be easily corrected, while colloidal stabilisers such as gums and alginates can be used for foam formation, but excessive use of these compounds can cause turbidity.

Lack of foam is usually due to a lack or excess of carbon dioxide in the beer, low malt concentration, excessive filtration, high levels of higher alcohols, too much use of anti-turbidity adjuncts or contamination of the drink with fat or oil.

5.3 Phenolic

Caused by the presence of chlorophenol or chemically similar molecules, it creates a medicinal flavour in beer. This compound is perceived at concentrations above 5 ppb (5 µg/L). It is easier to detect by sensory methods than by analytical methods.

This defect can be caused by: contaminating bacteria or wild yeasts; chlorinated water, water contaminated by insecticides, organic matter or certain types of algae; cleaning and hygiene products.

An alternative way of trying to avoid this would be to treat the water with activated charcoal.

5.4 Sediment

Sediment can be caused by the flocculation of colloids and by micro-organisms that cause cloudiness, or by the remains of filtration, such as cellulose fibres, activated charcoal or diatomaceous earth, by excessive use of foam stabiliser or by varnish (lacquer) on lids or cans.

5.5 Diacetyl

It imparts a butter flavour to the drink and can be considered a contaminant or a quality to the product. When concentrations are higher than 0.10 ppm (100 µg/L) the compound takes on an unpleasant flavour.

Diacetyl is a product generated by yeast metabolism during fermentation. At the beginning of this stage, diacetyl levels rise to 0.5 ppm, but after digestion of the fermentable sugars, the micro-organism itself metabolises the compound, reducing it to concentrations below 0.10 ppm. However, it can happen that an insufficient amount of yeast is used during fermentation, which will favour the reduction of the compound.

Another reason for the appearance of diacetyl is the contamination of the must by lactic acid bacteria of the genus *Pediococcus*. For this reason, it is important to aseptically clean the equipment and, if necessary, treat the yeast (when contaminated by bacteria) with acidic solutions in order to stop the infection.

5.6 Sulphurous to yeast

The sulphurous smell is also generated during the fermentation stage by the yeast, but is eliminated after clarification, filtration and pasteurisation of the liquid. When ascorbate is used as the only antioxidant, the beer develops a bread-like flavour, but if sulphur dioxide is used as an antioxidant, the smell of hydrogen sulphide will appear.

The factors that contribute to the appearance of these sulphurous odours are: Low oxygenation or poor boiling of the wort, poor yeast breeding and growth, storing the beer after bottling at high temperatures, exposing the beer to light (wavelengths below 550nm are the most damaging to the quality of the drink) and slow fermentation.

5.7 Old or oxidised

Flavour developed when beer deteriorates. The spoilage process begins as soon as the fermentation stage is finished. Over time, the undesirable compounds that are present in the drink (at very low levels) begin to develop and increase in concentration. Unlike spirits, which need ageing to refine their flavour, beer should be consumed as soon as it is ready.

The main actions to slow down the ageing of the drink are: Using quality raw materials and a suitable brewing process, storing bottled beer at low temperatures and pasteurising the drink, even minimally.

CHAPTER 6

BIBLIOGRAPHICAL REFERENCES

ABOUMRAD, Jean Pierre Cordeiro; BARCELLOS, Yvie Carolinne Medeiros. 82 f. **Analysis and Simulation of the Mustering and Fermentation Operations in the Beer Production Process.** 2015. Dissertation (Degree in Chemical Engineering) - Department of Chemical and Petroleum Engineering. Fluminense Federal University. Niterói. 2015.

AMBEV. **A tour through history.** About Beer: The long journey of beer. 2016. Available at: <https://www.ambev.com.br/conhecimentocervejeiro/pt-br/history-of-beer>. Accessed on: 10 July 2018.

AQUARONE, Eugênio; LIMA, Urgel de Almeida; BORZANI, Walter. **Foods and Beverages Produced by Fermentation.** 5ª ed. São Paulo. Blucher, 1993. 243 p.

BRAZILIAN BEER INDUSTRY ASSOCIATION. **Fun facts.** Cervbrasil. São Paulo, 2018. Available at: <http://www.cervbrasil.org.br/novo site/curiosidades/>. Accessed on: 12 July 2018.

BRAZILIAN BEER INDUSTRY ASSOCIATION. **Benefits of Beer.** Cervbrasil Beer & Health tab. Available at: < http://www.cervbrasil.org.br/paginas/index.php?page=benefícios-da-cerveia>. Accessed on: 23 March 2017.

BRAZILIAN BEER INDUSTRY ASSOCIATION. **2015 Yearbook.** Cervbrasil. São Paulo, 2015. 52 pages. Available at: < http://www.cervbrasil.org.br/arquivos/ANUARIO CB 2015 WEB.pdf>. Accessed on: 23 March 2017.

BJCP, Beer Judge Certification Programme. **Study guide for the BJCP beer exams.** 1998-2012. Available at: <https://www.bjcp.org/intl/Study- EN.pdf>. Accessed on: 01 June 2018.

BOBBIO, Florinda O.; BOBBIO Paulo A. **Introduction to Food Chemistry.** 3ª ed. São Paulo. Varela, 2003. 238 p.

BORZANI, Walter et al. **Industrial Biotechnology: Beers.** 2ª ed. São Paulo. Blucher, 2008. Vol. 4. 91-144 p.

BRAZIL, Decree No. 6871, of 4th June 2009. Regulates Law No. 8.918, of 14 July 1994, which provides for the standardisation, classification, registration, inspection, production and supervision of beverages. **Lex:** Decreed by the President of the Republic Luiz Inácio Lula da Silva, Brasília, 4 June 2009.

CORAZZA, Rodrigo Marcos. 40 f. **The recent expansion of craft breweries in the context of high concentration of the beer market in Brazil.** 2011. Dissertation (Degree in Economic Sciences) - Institute of Economics. UNICAMP. Campinas. 2011.

D'AVILA, Roseane Farias; LUVIELMO, Márcia de Mello; MENDONÇA, Carla Rosane Barboza; JANTZEN, Márcia Monks. **Adjuncts used in beer production: Characteristics and Applications.** 2012. Technological Studies in Engineering. Federal University of Pelotas and Federal University of Rio Grande do Sul. Volume 8, N.2, pages 60-68. Jul/Dec 2012.

FERRETI, Fernanda Luísa. 30 f. **The craft beer market and beer culture in Greater Florianópolis.** 2014. Dissertation (Degree in Journalism) - Centre for Communication and Expression CCE, Department of Journalism. UFSC. Florianópolis. 2014.

FRANCO, Bemadette D. Gombossy de Melo; LANDGRAF, Mariza. **Food Microbiology.** Iª ed. São Paulo. Atheneu, 2005. 182 p.

GUEIROS, F. J. **Carbohydrates.** USP: Institute of Chemistry. Available at: <http://www2.iq.usp.br/docente/fgueiros/Carboidratos alunos.pdf>. Accessed on: 15/05/2017.

MADRID, A.; CENZANO, L; VICENTE, J. M. **Manual of Food Industries.** Iª ed. São Paulo. Varela, 1996. 598 p.

MIYAMOTO, S. Sugars: Structure and Function. USP: Institute of Chemistry. Available at: <http://www2.iq.usp.br/docente/miyamoto/QBQ0105%20Enfermagem/Aula Carbohydrate Structure.pdf>. Accessed on 15/05/2017.

NIQUITO, Thomas Esteves. **Craft Beer Production Course.** Cia Paulista Pub. Rio Claro, 23 July 2016. Course given by craft brewer and beer sommelier Thomas E. Niquito.

OETTERER, Marília; REGITANO-D'ARCE, Marisa Apª Bismara; SPOTO, Marta Helena Fillet. **Fundamentals of Food Science and Technology.** Iª ed. Manole, 2006. 612 p.

PICCINI, Ana Rita; MORESCO, Cristiano; MUNHOS, Larissa. **Beer.** April 2002. <http://www.uffgs.br/alimentusl/feira/prcerea/cerveja/filtl.htm>. Accessed on: 09/07/2018.

ROSA, Natasha Aguiar; AFONSO Júlio Carlos. The Chemistry of Beer. **Química Nova na Escola,** São Paulo, Vol. 37, n°2, p. 98-105, May 2015.

SANTOS, Ana Cristina de Ávila. 66 f. **Monitoring Dissolved Oxygen in beer during the Fermentation and Maturation process.** 2003. Dissertation (Degree in Food Engineering). Catholic University of Goiás - UCG. Goiânia, 2003

SILVA, João Batista de Almeida e, Chap. 15: Beer. FILHO, Waldemar G. Venturi. **Beverage Technology: Beers.** Iª ed. São Paulo. Blucher, 2005. 347-382 p.

SILVA, Paulo Henrique Alves da; FARIA, Fernanda Carolina de. Evaluation of the intensity of bitterness and its active principle in beers of different characteristics and commercial brands. **Scielo.** Campinas, p. 902-906. Oct/Dec. 2008.

Available at: <www.scielo.br/pdi7cta/v28n4/a21v28n4.pdf>. Accessed on: 10 June

2018.

VENTURINI FILHO, W.G. 2005. **Beverage Technology:** Raw materials, processing, GMP/APPCC, Legislation, Market. São Paulo, 2005 Edgard Blucher, 550 p.

Printed by Books on Demand GmbH, Norderstedt / Germany